人类与千年洪水

Humans and the Floods Across Millennia

（汉英对照）

中国水利水电科学研究院
国际洪水管理大会 编著

科学普及出版社
·北 京·

图书在版编目（CIP）数据

人类与千年洪水：汉英对照 / 中国水利水电科学研究院，国际洪水管理大会编著 . -- 北京：科学普及出版社，2024. 10. -- ISBN 978-7-110-10869-7

Ⅰ. P426.616-49

中国国家版本馆 CIP 数据核字第 202450G8D5 号

策划编辑 李 睿
责任编辑 李 睿
图书装帧 瑞东国际
责任校对 焦 宁
责任印制 徐 飞

出　　版 科学普及出版社
发　　行 中国科学技术出版社有限公司
地　　址 北京市海淀区中关村南大街 16 号
邮　　编 100081
发行电话 010-62173865
传　　真 010-62173081
网　　址 http://www.cspbooks.com.cn

开　　本 787mm × 1092mm 1/16
字　　数 80 千字
印　　张 3.5
版　　次 2024 年 10 月第 1 版
印　　次 2024 年 10 月第 1 次印刷
印　　刷 北京博海升彩色印刷有限公司
书　　号 ISBN 978-7-110-10869-7/P · 247
定　　价 58.00 元

编 委 会

序

水是生命之源、万物之本。水在滋养哺育人类的同时，也塑造了人类的历史。纵观中华文明的发展史，一定程度上也是一部与洪涝、干旱作斗争的历史。从古至今，中华民族一直在与水相伴、相争、相和中发展。在漫长的历史长河中，洪水一直是威胁人类生存的主要自然灾害，经常被比喻为“洪水猛兽”。洪水给人类造成了危害，也重塑了地形地貌，一直深刻影响着人类文明的进程。面对洪水，人类从未放弃过努力，从最初的被动应对到后来的主动防御，再到如今追求与洪水共生共荣，这一历程深刻反映了人类与自然环境关系的演变和进步。

党的十八大以来，习近平总书记多次对科普宣传教育工作作出重要指示批示，强调“科技创新、科学普及是实现创新发展的两翼，要把科学普及放在与科技创新同等重要的位置”。为加强我国的防洪抗旱减灾科普知识宣传工作，拉近防洪抗旱减灾与社会公众的距离，提高民众的应急避险和自救互救能力，这本《人类与千年洪水》应运而生。本书希望通过寓教于乐的方式普及洪水的科学知识，提升读者对水患灾害的认知和应对能力。

本书依托防洪抗旱减灾科普教育基地，由中国水利水电科学研究院张诚等专家领衔撰写。在书中，读者将跟随小水滴家族的足迹，穿越历史长河，见证都江堰、苏公堤、荷兰围海大堤、长江三峡等世界伟大的水利工程，感受不同时代、不同地域的人们在洪水面前展现的智慧与勇气，探索洪水与人类共存的千年智慧。希望每一位读者都能在知识与趣味的交织中获得启发，为保护我们共同的家园贡献自己的力量。

是为序。

中国工程院院士

2024 年 9 月 1 日

Preface

Water is the source of life and the origin of all things. Nourishing and sustaining humanity, water has shaped human history. The history of Chinese civilization is partly the history of combating floods and droughts. Since ancient times, the Chinese people have lived, struggled, and harmonized with water. In the long course of history, flooding has remained one of the major types of natural disasters threatening human survival, often metaphorically referred to as "raging floods and roaring beasts". Floods have caused harm, reshaped terrains, and profoundly influenced the course of human civilization. In facing floods, human beings never given up – experiencing a change from passive coping to proactive defense and now striving for coexistence and harmony with floods. This is a journey marking the evolution and progress of the relationship between human and nature.

Since the 18th National Congress of the Communist Party of China, General Secretary Xi Jinping has repeatedly emphasized the importance of science popularization. He has pointed out that "scientific and technological innovation and dissemination of scientific knowledge are the two wings to propel our innovation-driven development. The latter should be considered as important as the former." The book, *Humans and the Floods Across Millennia*, is created to enhance the awareness of flood control, drought relief, and their disaster reduction in China, bringing these concepts closer to the general public and improve their disaster preparedness and self-rescue capabilities. It is designed to share scientific knowledge about floods in an engaging and educational manner and to help young readers better understand and prepare for the impacts of water-related disasters.

The book has been authored by experts led by Zhang Cheng from the China Institute of Water Resources and Hydropower Research. Readers will follow the journey of Waterdrop family through history, and glimpse the great hydraulic civilizations around the world, such as Dujiangyan, the Su Causeway, the Netherlands Dikes and Three Gorges Dam. Readers will be impressed by the wisdom and courage displayed by people from different time periods and different parts of the world in confronting floods and exploring their own ways of living with floods. Hopefully we will all have fun reading this interesting book and do our bit for protecting our planet.

CAE Academician
Wanghao
September 1, 2024

Living with Floods: Learning, Understanding, and Staying Safe

Floods are a natural part of our world – they happen when water from rivers, oceans, or heavy rains spills over and covers land that's usually dry. You might have seen floods on TV or in stories, with water rushing through towns or filling up fields. Floods can be slow and gentle, rising gradually, or they can arrive suddenly, like a "flash flood," sweeping over everything in their path. Because floods are so powerful, it's important to understand how they happen, why they're dangerous, and, most of all, how we can be prepared to stay safe.

This book will help children learn all about floods, so they can feel more prepared if they ever experience one. The book explores where floods come from and the different ways they start. Floods might happen when there's a lot of rain that can't soak into the ground, or when rivers overflow their banks after a storm. Sometimes, floods come from big ocean waves called "storm surges," which can push water far onto land. Even melting snow or ice can cause rivers to swell, leading to flooding. With so many ways for water to spill out of its usual places, floods are an important part of nature that we all need to understand.

Around the world, people have come up with smart ways to protect their homes, families, and communities from floods. The book provides some examples, like in the Netherlands, much of the land is below sea level, so they build big, strong walls called "dikes" to keep the ocean out. In Venice, Italy, they use special gates to hold back water during high tides so the streets don't flood. And in Japan, huge underground tunnels help direct extra water safely out of the city when rivers overflow. These different ideas show how people all over the world have learned to live with floods by being prepared.

In this book, the kids will also discover ways to protect themself and their family if a flood happens where they live. Having a safety plan is important! This means knowing where to go if you need to leave your home, and packing an emergency kit with things like water, food, a flashlight, and a blanket. Listening to warnings and being ready can make a big difference.

By reading this book, the kids will understand the power of water and how people work together to stay safe during floods. They will learn how being prepared can help keep them, their family, and their community safe. With the knowledge they gain here, they'll be ready to face the possibility of a flood with confidence and be a part of keeping their community safe too!

University of Western Ontario
ICFM Chairperson
Fellow of Canadian Academy of Engineering
Fellow of Royal Society of Canada
Prof. Slobodan P. Simonovic

与洪水共生：掌握知识与应急避险

洪水是我们这个世界与生俱来的一部分——当来自河流、海洋或暴雨的水漫溢并泛滥于原本干燥的土地时，就会发生洪水。或许你曾经在电视上或故事中看到过洪水，湍急的水流穿过城镇或淹没田野。洪水可能是缓慢而温和的，逐渐上涨；也可能突然袭来，如突发性洪水，席卷沿途的一切。洪水的威力极大，因此，有必要了解它们的成因、危险性以及如何做好应急避险准备。

本书将帮助小读者全面地认识洪水，以便在遇到洪水时能够更加从容地应对。本书探讨了洪水的来源以及不同的成因。降雨量过大、未能渗入地下时，或者在暴风雨过后河水漫过堤岸时，都可能发生洪水。有时，洪水来自被称为风暴潮的巨大海浪，这种海浪会将水推到远离海岸的地方。甚至冰雪融化也会导致河流水位上涨，从而引发洪水。导致洪水冲决泛滥的原因数不胜数，因此洪水是我们需要了解的一种重要的自然现象。

在世界各地，人们已经想出了许多聪明的方法来保护家园、家庭和社区免受洪水的侵害。书中列举了一些例子。例如，在荷兰，大部分土地低于海平面，因此人们建造了坚固的大型堤坝来阻挡海水。在意大利的威尼斯，人们使用特殊的防洪闸门，在涨潮时阻止海水涌入街道。在日本，每当河流暴涨时，巨型地下隧道将多余的水安全地排出城市。这些不同的例子展示出世界各地的人们都已经学会通过防患于未然来与洪水共存。

在本书中，小读者们还将学习在洪水发生时应该如何保护自己和家人。提前制订安全计划至关重要！这意味着知道在疏散时应该前往何处避险，并准备一个应急包，其中包括水、食物、手电筒和毯子。留意警报并随时做好准备可以起到很大的作用。

通过阅读本书，小朋友们将了解水的威力，了解人们如何在洪水期间共同努力，保持安全。他们将认识到，防患于未然有助于确保自己、家人和社区的安全。凭借从本书中学到的知识，他们将能够充满自信地应对可能发生的洪水，并且为社区的安全做出自己的贡献！

国际洪水管理大会主席

加拿大工程院院士

加拿大科学院院士

加拿大西安大略大学教授

斯洛博丹·西蒙诺维奇

目录 CONTENTS

性别：男
Gender: Boy
年龄：6 岁
Age: 6
性格：好奇心强，活泼调皮
Personality: Curious, energetic, and a bit mischievous
口头禅：咦，为什么会这样呢？
Catchphrase: Huh? Why is it like this?

大家好，我是小水滴。我最喜欢探索未知的事物啦！让我们一起来学习洪水的知识吧！
Hi, everyone! I'm Little Waterdrop. I love discovering new things and solving mysteries. Let's dive into the fascinating world of floods together!

各位，我是水滴爸爸。本次旅行的引领者和超级顾问，就是我！让我们一起探索水的奇妙吧！
Hi there! I'm Waterdrop Dad, your guide and expert for this exciting journey. Let's explore the wonders of water together!

我是水滴妹妹。我正在学习如何成为一个更好的探险家！和我一起揭开洪水的秘密吧！
Hello! I'm Waterdrop Sister. I'm learning how to be a brave explorer. Join me as I uncover the secrets of floods!

性别：女
Gender: Girl
年龄：4 岁
Age: 4
性格：俏皮可爱，胆子小
Personality: Playful, sweet, but a little shy
口头禅：天啊！这是怎么回事？
Catchphrase: Oh no! What's going on?

性别：男
Gender: Man
年龄：35 岁
Age: 35
性格：博学多才，温和风趣
Personality: Knowledgeable, kind, and funny
口头禅：问得好！
Catchphrase: Great question!

他已经忙得说不出话来了……
He was so busy he couldn't even respond…
咦，水滴爸爸在干什么呢？
Huh? What's Dad up to?
孩子们，这是我送给你们的礼物——时空飞船。
Kids, this is my gift to you—a Time-Space Ship!
哇！
Wow!
这艘飞船能带我们穿越时空吗？
Wow! Can this ship really take us through time and space?
必须可以！
Of course it can!
我们想去哪里就能去哪里吗？
Can we go anywhere we want?
当然！
Absolutely!
太棒了！
This is amazing!

孩子们，一起出发吧！

Kids, Let's Set Off on Adventures!

炎炎夏日，小水滴一家迎来了令人期待的暑假，这一天，小水滴和水滴妹妹乘坐着水滴爸爸发明的时空飞船，开启了一场别开生面的亲子旅行。

In the scorching summer, the Waterdrop family welcomed an eagerly anticipated vacation. Little Waterdrop, his sister, and their dad boarded a Time-Space Ship to embark on a unique parent-child adventure.

飞着飞着，时空飞船来到了一个村庄的上方。此时，暴涨的河水如同一群咆哮的野兽，气势汹汹地向前奔涌，肆无忌惮地冲垮了河堤，淹没了农田，吞噬了道路。

As the ship soared through the skies, it soon arrived above a village. But something was wrong. Below, the river had turned wild, its waters swelling and thrashing like angry monsters. The raging current smashed through riverbanks, swallowed up farmlands, and devoured roads without hesitation.

孩子们，这是“洪水”。
Kids, what you're seeing is called a “flood.”

太吓人了！
That's so scary!

这么大的水，也是我们水滴家族的成员吗？
Whoa... all that water—is it part of our Waterdrop family too?

水滴家族话科普

The Waterdrop Family's Science Talk

水滴爸爸，什么是洪水呀？
Daddy, what is a flood?

洪水，指因大雨、暴雨或持续降雨使低洼地区淹没、渍水的现象。
A flood refers to the phenomenon where low-lying areas are submerged or waterlogged due to heavy rain, torrential downpours, or continuous rainfall.

我懂了，水是生命之源，但也不是越多越好！
Oh, I see! Water is essential for life, but too much of it can be a problem!

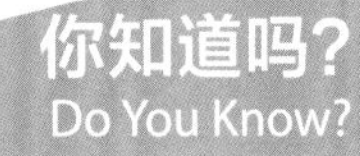

你知道吗？
Do You Know?

洪水会损害农作物，破坏建筑物和各类设施，造成人员伤亡和财产损失，污染水源，传播疾病，使交通、通信中断，等等。

Floods can cause serious damage: They ruin crops and destroy buildings and other structures. They can harm people, cause property loss, and contaminate water supplies. Floods may also spread diseases and disrupt transportation and communication.

可怕的史前“猛兽”

Terrifying “Beasts” of Prehistoric Times

一不小心，水滴妹妹碰到了穿越按钮，时空飞船瞬间来到了遥远的史前时代。谁知，这里正遭受着史前大洪水这个“猛兽”的侵袭！

Waterdrop Sister accidentally pressed the time travel button on the Time-Space Ship. In the blink of an eye, they were whisked away to the distant prehistoric era. But what awaited them was no peaceful scene—the land was being ravaged by a colossal prehistoric flood, a terrifying “beast” of nature!

高山在洪涛中颤抖，陆地在巨变中呻吟……紧急时刻，一艘能为生灵提供庇护的诺亚方舟出现了！

Amid the roaring torrents, mountains shook, and the land seemed to cry out in pain as it shifted and changed. Just when all hope seemed lost, a giant ark appeared on the horizon—a refuge to protect life from the raging waters.

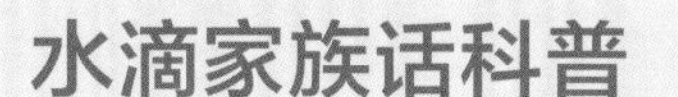

水滴家族话科普

The Waterdrop Family's Science Talk

水滴爸爸，洪水自古以来就给人类带来了很多灾难吗？
Dad, have floods caused a lot of disasters throughout history?

是啊。据史料记载，从公元前 206 年到公元 1949 年，我国发生了 1029 次大水灾，几乎每两年就有一次。在西亚的底格里斯 - 幼发拉底河流域和非洲的尼罗河流域，关于洪涝的记载可以追溯到公元前 40 世纪。
They certainly have. According to records, there were 1029 major floods in China from 206 BC to 1949 AD—almost one every two years. In the Tigris-Euphrates region of West Asia and along the Nile River in Africa, stories of flooding go all the way back to the 40th century BC!

看来，在人类的历史中，洪水一直都存在啊！
It seems that floods have been around for as long as humans can remember!

世界各大古文明都有关于史前大洪水的记载。人类表达出对洪水的恐惧，并将其视作神祇、用作图腾。

Ancient civilizations all over the world recorded stories of great prehistoric floods. People feared them so much that they turned them into gods or used them as symbols in their cultures.

洪水的多张“面孔”
The Many “Faces” of Floods

亲眼见证了惊心动魄的史前大洪水，小水滴和水滴妹妹久久不能平静！他们对洪水“猛兽”产生了极大的好奇。

After witnessing the breathtaking prehistoric flood, Little Waterdrop and Waterdrop Sister couldn't stop thinking about it. They were full of curiosity about this mighty “beast” called a flood.

孩子们，洪水可不只有一张“面孔”。
Kids, did you know floods don't just have one “face” ?

真的吗？它还会“变脸”吗？
Really? Do they have different “faces” ?

快带我们去见见洪水的不同“面孔”吧！
Let's go explore and see all the different “faces” of floods!

水滴家族话科普
The Waterdrop Family's Science Talk

洪水究竟有哪些“面孔”呢？
What are the different “faces” of floods?

暴雨洪水、山洪、风暴潮、融雪洪水、冰凌洪水、溃坝洪水等，这些都是洪水的类型。
Floods come in many forms, including: flash floods, mountain torrents, storm surges, snowmelt floods, ice jam floods, dam-break floods, etc.

洪水的“面孔”真是多种多样呀！
Floods really do come in so many faces!

在水滴爸爸的带领下，时空飞船在不同时空之间飞快地穿梭。瞧，不同类型的洪水逐一出现了！

Under Waterdrop Dad's lead, the Time-Space Ship zoomed through different eras and places. Look! Various types of floods were appearing right before our eyes!

暴雨洪水 Flash Floods

由强度较大的降雨形成，是最常见的洪水。

Flash floods are caused by heavy rainfall and are the most common type of flood.

山洪 Mountain Torrents

由山区溪流沟涧地带的暴雨引发。

Mountain torrents are caused by heavy rainfall in the streams and ravines of mountainous areas.

风暴潮 Storm Surges

由强烈大气扰动而引起的海面异常升降的现象。

Storm surges are a phenomenon caused by intense atmospheric disturbances that lead to abnormal rises and falls in sea levels.

融雪洪水 Snowmelt Floods

由冬季积雪或冰川在春夏季节随气温升高融化而形成的洪水。

Snowmelt floods are floods caused by the melting of accumulated snow or glaciers during the spring and summer seasons as temperatures rise.

冰凌洪水 Ice Jam Floods

河流中因冰凌阻塞和河道内蓄冰、蓄水量的突然释放而引起的显著涨水。

Ice jam floods are significant rises in water levels caused by ice blockages in rivers and the sudden release of accumulated ice and water within the river channel.

溃坝洪水 Dam-Break Floods

由于堤坝或其他挡水建筑物瞬时溃决，水体突泄而产生的洪水。

Dam-break floods are floods caused by the sudden failure or breach of a dam or other water-retaining structures, resulting in a rapid release of water.

洪水和古文明有什么关系？

Floods and Ancient Civilizations: What’s the Connection?

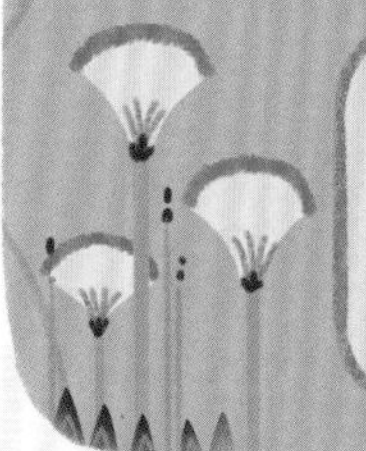

刚刚，兄妹俩见识了不同类型的洪水。接下来，水滴爸爸又会带领他们前往哪里旅行呢？

After seeing so many different kinds of floods, Little Waterdrop and Waterdrop Sister were curious about what comes next. Where would Waterdrop Dad take them on their next adventure?

时空飞船来到了古代的尼罗河谷。看，这里有肥沃的土地，有小麦、大麦等农作物，还有高大挺拔的棕榈树。

The Time-Space Ship touched down in the ancient Nile River Valley. Look at this beautiful place! The land was rich and fertile, with crops like wheat and barley growing tall, and palm trees stretching toward the sky.

水滴家族话科普
The Waterdrop Family's Science Talk

水滴爸爸，我听说太阳历的发明和洪水有关？
Dad, I heard the invention of the Solar Calendar has something to do with floods?

没错。为记录洪水泛滥的时间，古埃及人通过天文观察，发现365天形成一“年”，从而发明了人类历史上最早的太阳历。
That's right! To keep track of the flooding seasons, the ancient Egyptians observed the stars. They discovered that a year is made up of 365 days, and that's how they invented the first solar calendar in history.

哇，没想到，洪水还和古文明有关呢！
Wow, I didn't know floods were connected to ancient civilizations!

太阳历
Solar Calendar

你知道吗？
Do You Know?

世界上第一部成文法典《汉穆拉比法典》中，记载了古巴比伦保护农田的堤堰。在当时，人们就懂得修筑防洪设施，逐渐开启了与洪水共生的探索。

In the world's first written legal code, *the Code of Hammurabi*, there's a record of ancient Babylonians building levees to protect their farmland. People back then already understood the importance of flood defenses, which was the beginning of humanity's journey to live in harmony with floods.

就不信治不了你！
I don't believe we can't control you!

飞船驶离古代的尼罗河，小水滴心想："面对洪涝灾害，古代人是如何应对的呢？"这时，水滴爸爸扭转方向盘，带大家来到了中国的上古时期。

As the ship zoomed away from the ancient Nile River, little Waterdrop wondered, "How did people in ancient times deal with floods?" Just then, Waterdrop Dad turned the wheel, guiding them to ancient China.

① 三皇五帝时期，鲧和禹父子二人受命治理黄河的水患。

During the era of the Three Sovereigns and Five Emperors, Gun and his son Yu were tasked with the responsibility of controlling the flooding of the Yellow River.

② 鲧筑三仞之城围堵洪水，但水却越淹越高。

Gun built a massive city to hold back the floodwaters, but the water just kept rising higher and higher.

③ 鲧的儿子禹把淤塞的川流疏通，将洪水引入河道、洼地或湖泊，成功治理了水患！

But Yu, Gun's son, had a clever plan. He cleared out the blocked rivers and carefully guided the floodwaters into channels, lowlands, and lakes. Finally, he succeeded in controlling the floods!

水滴家族茶话会

The Waterdrop Family Tea Party

大禹真聪明，治水方法科学又创新！

Yu was so clever! His methods for controlling the floodwaters were both scientific and innovative!

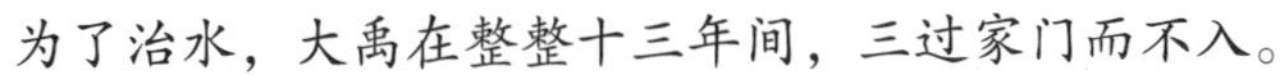

为了治水，大禹在整整十三年间，三过家门而不入。

To control the floods, Yu worked tirelessly for thirteen years, passing his home three times without ever going in.

这也太大公无私了吧？真令人敬佩！

That's so selfless! What an amazing person!

共工怒触不周山的神话传说

The Legend of Gong Gong and Mount Buzhou

据《淮南子》记载，水神共工曾与颛顼（zhuān xū）争夺部落天帝之位，但失败了。恼羞成怒的共工撞断了不周山，导致江河积水泥沙都朝东南角流去。古人认为这就是远古时期大洪水肆虐的原因之一。

According to *the Huainanzi*, the water god Gong Gong once tried to take the throne of the Heavenly Emperor from Zhuanxu but lost. Furious, Gong Gong angrily struck Mount Buzhou, breaking it apart. This caused the rivers to flood and the mud to rush towards the southeast. This was believed by ancients to be one of the reasons for the great floods.

就地取材的智慧

Smart Solutions: Using What's Around Us

从上古时期离开，水滴家族乘坐着时空飞船，来到了古代农耕时期。这次，他们会有怎样的新发现呢？

Leaving the ancient times behind, the Waterdrop family boarded their Time-Space Ship and traveled to the era of farming civilizations. What exciting new discoveries will they make this time?

快看！这里有一群忙碌的古代人。
Look over there! A group of people is hard at work.

他们在做什么呀？
What are they doing?

他们在修建堤坝呢！
They're building a dam!

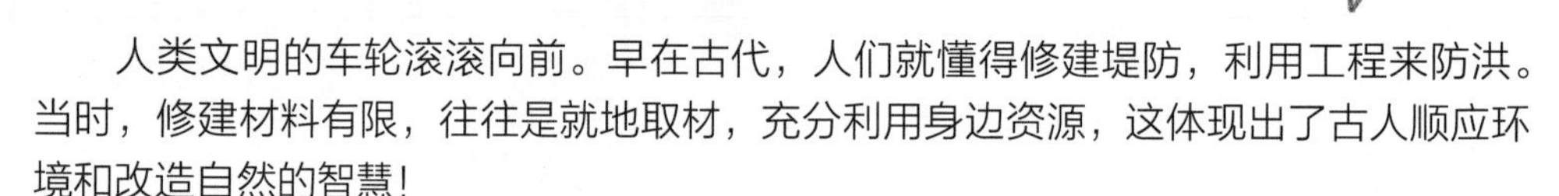

人类文明的车轮滚滚向前。早在古代，人们就懂得修建堤防，利用工程来防洪。当时，修建材料有限，往往是就地取材，充分利用身边资源，这体现出了古人顺应环境和改造自然的智慧！

As civilization moved forward, ancient people learned how to build levees to keep floodwaters at bay. With limited materials back then, they had to get creative, using whatever was around them. This shows how the ancient people worked with nature, finding smart ways to adapt and transform their surroundings!

天然材料：土、石、木

Natural materials: soil, stone, wood

水滴家族话科普
The Waterdrop Family's Science Talk

水滴爸爸，堤防有什么作用呢？
Dad, what does a levee do?

堤防是沿河、沿湖或沿海的防水构筑体，它可以约束水流，防御洪水，还能够防御风浪和海潮。
A levee is a structure built along rivers, lakes, or coasts to hold back water. It helps control the flow of water, prevent floods, and protect against waves and tides.

古人就地取材建堤防，真棒！
Wow, the ancient people were so smart to use what they had around them to build levees!

你知道吗？
Do You Know?

除了堤防，水库也有防洪的作用哦！

水库是拦蓄洪水的镇水重器，同时也能调节水资源的分配，灌溉农田，还可以发电，带来清洁能源。

Besides levees, reservoirs are also key to controlling floods!

They don't just hold back floodwaters, but also help manage water, irrigate crops, and even create clean energy for our world.

巧夺天工都江堰

A Marvel of Engineering: Dujiangyan Irrigation System

时空飞船再次启动，水滴爸爸带领大家来到战国时期的巴山蜀水间。看，这里有一座古老的大型水利工程—— 都江堰。

The Time-Space Ship hummed to life again. This time, Waterdrop Dad led the two kids back to the Warring States period, to the lush lands of Sichuan.Look! That's the Dujiangyan Irrigation System, an ancient masterpiece of hydraulic engineering!

汹涌的岷江水经过都江堰的调节后，平稳地流向水渠，灌溉万顷农田。都江堰的修建，使旱涝无常的成都平原变成了沃野千里的“天府之国”。

Thanks to the Dujiangyan system, the mighty Minjiang River is carefully controlled, guiding its waters smoothly into canals that water fields for miles. This amazing project turned the Chengdu Plain—once flooded or dried out by unpredictable weather—into the "Land of Abundance," a place where crops grow in plenty!

水滴家族话科普

The Waterdrop Family's Science Talk

水滴爸爸，都江堰的修建在当时也是就地取材吗？

Dad, was the Dujiangyan Irrigation System built using local materials too?

问得好！没错，都江堰工程以竹、木、卵石为主，形成了独具特色的四大传统水工技术——竹笼、杩槎（mǎchá）、羊圈、干砌卵石。

Great question! Yes, Dujiangyan was mainly built with bamboo, wood, and pebbles. It also uses four clever, traditional techniques: bamboo cages, wooden tripods, stockades, and dry masonry with pebbles.

竹笼
bamboo cages

杩槎
wooden tripods

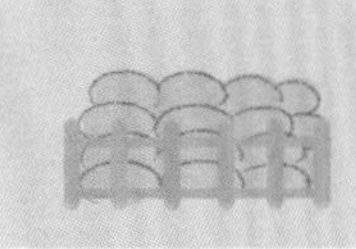

羊圈
stockades

干砌卵石
dry masonry with pebbles

哇，都江堰真是古代的水利工程奇迹呀！

Wow! The Dujiangyan system is truly an ancient marvel of water engineering!

伟大的治水英雄——李冰父子

The Great Heroes of Flood Control — Li Bing and His Son

2200 多年前，蜀郡守李冰和他的儿子率众兴建这座能分洪减灾、灌溉农田、行舟走船的无坝引水工程——都江堰。直到今天，这项工程仍在造福人民，被誉为“活的水利博物馆”。

Over 2200 years ago, Li Bing, the Governor of Shu, and his son led the construction of the Dujiangyan Irrigation System. This amazing system can control floods, irrigate farmlands, and even let boats pass through! Even today, it continues to benefit the people and is known as the "Living Water Museum".

美丽的苏公堤
The Beautiful Su Causeway

返回飞船，小水滴和水滴妹妹愉快地玩了起来。玩闹间，穿越按钮被按下！时空迅速流转，他们来到了北宋年间的西湖。

Back on the Time-Space Ship, Little Waterdrop and his sister were happily playing. Suddenly, the time-travel button was pressed by accident! In a flash, they were zooming through time, landing in the West Lake during the Northern Song Dynasty.

苏公堤宛如一条蜿蜒的玉带，横卧于西湖之上，为西湖平添了一抹宁静而美丽的色彩。

The Su Causeway stretched across the lake like a winding ribbon of jade, adding a peaceful, shimmering beauty to the scene.

水滴家族话科普

The Waterdrop Family's Science Talk

水滴爸爸，苏公堤是怎样建成的呢？

Dad, how was the Su Causeway built?

问得好！苏公堤是宋代文人苏轼任杭州知府时修建的，堤身用疏浚西湖时挖出的淤泥和葑（fēng）草，再掺入沙石加固构筑而成，解决了泥沙淤积之患。

Great question! The Su Causeway was built by Su Shi, a famous poet and governor of Hangzhou, during the Song Dynasty. He used silt and fenchal grass dug up from the West Lake and added sand and stones to make it strong. This clever design helped stop the buildup of mud in the lake.

哇，这也是一座就地取材的堤坝！

Wow, it is also a dam made with materials from right around here!

“苏堤春晓”是什么？

What's "Spring Dawn at Su Causeway"?

苏公堤也称“苏堤”。春日里，苏堤上杨柳夹岸、艳桃灼灼，湖波如镜，意境动人，故称“苏堤春晓”。南宋时，苏堤春晓被列为西湖十景之首。

"Spring Dawn at Su Causeway" is one of the most famous views of West Lake. In the spring, the willows gently sway along the shores, the peach blossoms bloom brightly, and the lake's surface is so still it looks like a mirror. The scene is so serene and beautiful, it was named 'Spring Dawn at Su Causeway.' In the Southern Song Dynasty, it was regarded as the most enchanting of all the Ten Views of the West Lake.

荷兰的围海大堤

The Netherlands' Coastal Dikes

依依不舍地告别苏公堤，水滴家族踏上了新的旅程。他们来到“低洼之国”荷兰，看到了壮丽的围海大堤。

Reluctantly leaving Su Causeway behind, the Waterdrop family set off on a new adventure. Their next stop was the Netherlands, a country famous for its low-lying lands. There, they saw the impressive sea dikes, towering proudly against the sea.

水滴爸爸给兄妹俩讲起了荷兰围堤防洪、拦截水患的故事。

Waterdrop Dad then began telling the siblings a story about how the Netherlands built embankments to control floods and prevent water disasters.

①1287 年，荷兰遭遇了圣卢西亚洪水，5 万多人遇难。

In 1287, the Netherlands was struck by the catastrophic Saint Lucia Flood, which tragically claimed the lives of over 50000 people.

②洪水过后，荷兰人在新生成的须德海周围重新修筑海堤，用木桩及枝条编成阻波栅，围出淤积区，以防海水进一步倒灌。

After the disaster, the Dutch people rebuilt seawalls around the newly formed Zuiderzee. They used wooden stakes and branches to create wave barriers, enclosing areas of silt to prevent the sea from flooding the land further.

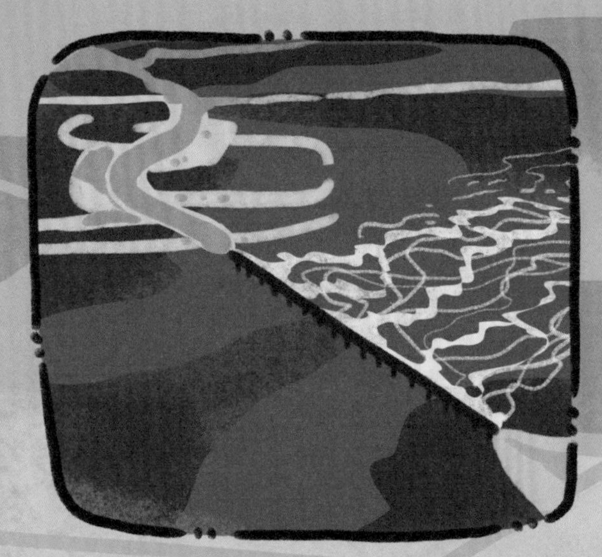

③早期的海堤系统为未来的三角洲工程奠定了基础。

This early seawall system laid the groundwork for the famous Delta Works that came later.

三角洲工程是什么？
What is the Delta Works?

荷兰人在鹿特丹以南的海湾之间修建的一系列水坝、防洪坝被称为三角洲工程。

The Delta Works is a series of dams and flood control barriers built by the Netherlands between the bays south of Rotterdam to protect the land from the sea.

建坝“神器”——混凝土

The Ultimate Dam Builder: Concrete

穿越茫茫的星际尘埃，水滴家族进入工业文明时代，踏上一片充满工业气息的土地。看，工业革命带来了空前的科技革新和飞跃！

Traveling through endless interstellar dust, the Waterdrop family arrived in the age of industrial civilization. They stepped onto a land buzzing with the energy of invention and progress. Look! The Industrial Revolution was bringing remarkable breakthroughs and incredible advancements in technology!

各位好！我是英国工匠约瑟夫·阿斯丁。1824 年，我发明了波特兰水泥。

Hello, everyone! I'm Joseph Aspdin, a British craftsman. Back in 1824, I invented Portland cement.

嗨，我是法国园丁约瑟夫·莫尼尔。1849 年，我试着在钢筋上浇筑混凝土，从而制成了钢筋混凝土。

Bonjour! I'm Joseph Monier, a French gardener. In 1849, I experimented by pouring concrete over steel bars, and voilà—reinforced concrete was born!

用混凝土建成的水坝拔地而起，抵挡住一波波洪水肆虐。于是世界各地都不约而同地掀起了一阵建坝高潮！

With concrete, dams began rising tall and strong, standing firm against relentless waves and floods. This revolutionary material sparked a global dam-building craze!

水滴家族话科普

The Waterdrop Family's Science Talk

水滴爸爸，用混凝土建水坝有什么优势呢？
Dad, why is concrete such a good material for building dams?

和木材、砖瓦等材料相比，混凝土强度更高，可以在较短时间内建造出更结实的水坝哦！
Great question! Compared to materials like wood or bricks, concrete is much stronger. It lets us build sturdier dams in less time!

原来，混凝土是建坝“神器”呀！
Wow, so concrete is truly the ultimate dam builder!

混凝土的出现是人类防御洪水，乃至利用洪水的福音。人类与洪水的共处也迎来了全新的时代——工业时代。

Concrete didn't just help humanity defend against floods—it also made it possible to harness their power. Its invention marked the beginning of a new chapter in human history: the Industrial Age, where people and water learned to work together like never before.

胡佛水坝，我们来啦！

The Mighty Hoover Dam, Here We Come!

见兄妹俩对混凝土水坝饶有兴趣，水滴爸爸笑着说：“我们去看看世界上第一座现代混凝土水坝吧！”

话音刚落，飞船穿越到了 20 世纪 30 年代的美国……

Seeing how fascinated the siblings were with concrete dams, Waterdrop Dad smiled and said, "How about we visit the world's first modern concrete dam?" Before they could respond, the Time-Space Ship whisked them away to 1930s America...

我认出来了，这里是美国科罗拉多河的黑峡！

I know where we are—this is the Black Canyon on the Colorado River!

哇，好壮观的水坝！

Wow! What an incredible dam!

对，这就是大名鼎鼎的胡佛水坝！

You've got it! This is the legendary Hoover Dam!

水滴家族话科普
The Waterdrop Family's Science Talk

水滴爸爸，胡佛水坝有何过人之处？
Dad, what makes the Hoover Dam so special?

问得好！大体积混凝土高坝筑坝技术，在它这里得到了创新性的发展。它在防洪灌溉、水力发电及航运等多方面发挥了巨大作用，还间接孕育了新兴城市拉斯维加斯呢！
Great question! The Hoover Dam revolutionized high-dam construction with large-volume concrete technology. It's a powerhouse for flood control, irrigation, hydropower, and shipping. And guess what? It even helped the rise of a city you might know—Las Vegas!

胡佛水坝可真厉害！
The Hoover Dam is incredible!

你知道吗？
Do You Know?

胡佛水坝的混凝土浇筑量达 300 多万立方米，厚度 200 米，犹如一把利铲直插峡谷。

The Hoover Dam contains over 3 million cubic meters of concrete and stands 200 meters thick! Its towering structure cuts through the canyon like a giant blade—a true marvel of engineering.

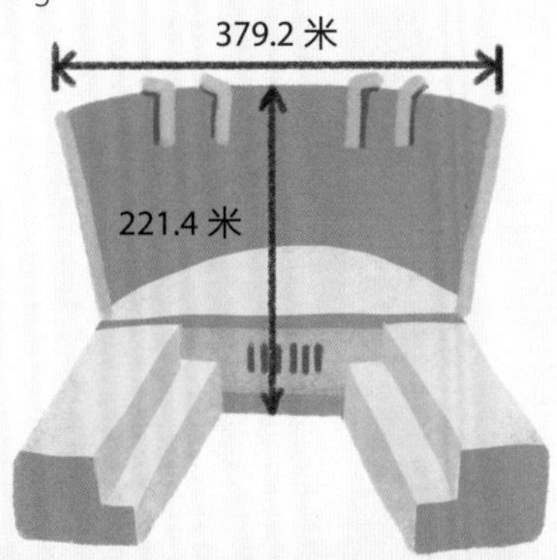

大国重器——三峡大坝

A National Wonder: The Three Gorges Dam

看完胡佛水坝，小水滴和水滴妹妹意犹未尽地返回时空飞船。突然，飞船快速转动，摇摆不定！

After their incredible visit to the Hoover Dam, Little Waterdrop and his sister hopped back aboard the Time-Space Ship, their excitement still bubbling. But suddenly— WHOOSH! The ship began to spin and wobble uncontrollably!

眨眼间，水滴家族来到了中国的湖北省宜昌市，眼前出现了壮丽的三峡大坝。

In the blink of an eye, the Waterdrop family arrived in Yichang City, Hubei Province, China, where they were greeted by the magnificent sight of the Three Gorges Dam.

水滴家族话科普
The Waterdrop Family's Science Talk

水滴爸爸，三峡大坝都有哪些作用？
Dad, what does the Three Gorges Dam do?

三峡大坝能够做到拦洪、削峰、错峰一气呵成，将防洪、蓄水、发电有力结合，大大造福了长江中下游百姓。
The Three Gorges Dam is like a superhero for the Yangtze River. It manages flood control, reduces peak water flows, and stores water, all while generating electricity. It's a huge help to the people living along the middle and lower parts of the Yangtze River.

哇，真了不起！
Wow, that's incredible!

你知道吗？
Do You Know?

三峡大坝是当今世界规模最大的混凝土重力坝。目前，全球注册的大坝数量超过 5.8 万座，为防御洪水发挥了重要的作用。

The Three Gorges Dam is the largest concrete gravity dam in the world. Globally, over 58000 dams have been built, serving as crucial barriers to protect against floods.

大坝高度：181 米；大坝长度：2335 米；混凝土总量：1700 万立方米

Dam height: 181m; Dam length: 2335m; Total concrete volume: 17 million m^3

隐蔽的地下宫殿
A Hidden Underground Temple

离开三峡大坝，时空飞船搭载着水滴家族前往日本东京。谁知，到达目的地之后，飞船竟然钻进了地下！

After leaving the Three Gorges Dam, the Time-Space Ship whisked the Waterdrop family off to Tokyo, Japan. But to everyone's surprise, the ship dove underground as soon as they arrived!

水滴家族话科普
The Waterdrop Family's Science Talk

水滴爸爸，人们为什么要修建这座地下宫殿呢？
Dad, why was this Underground Temple built?

问得好！这个地下宫殿拥有五个大立坑、四台蓄水池和大型水泵，是一个地下蓄水分洪系统，既能储水，也能排水。当水储存到一定量时，水泵会被打开，短短几秒就能把积水抽空。
Great question! This Underground Temple is actually a special flood control system. It's made up of five huge shafts, four large water reservoirs, and powerful pumps. It can store water and also drain it when needed. When the water reaches a certain level, the pumps kick in and drain the water in just a few seconds.

原来，这就是东京应对水灾的秘密武器呀！
So this is Tokyo's secret weapon against the floods!

你知道吗？
Do You Know?

英国伦敦也拥有出色的地下防洪工程，即地下水道系统，能有效地防止伦敦城市内涝。

London also has a fantastic underground flood control system—the city's sewer system. It helps keep the streets dry and prevents floods during heavy rains.

这是深藏于地下约 50 米处的东京圈排水系统，内有一条全长 6.3 千米、直径 10.6 米的巨型隧道。隧道里，一根根巨大的混凝土立柱如通天巨塔般巍然耸立，令人惊叹不已！

This system is part of Tokyo's Metropolitan Area Outer Underground Discharge Channel, buried about 50 meters underground. It includes a giant 6.3-kilometer-long tunnel with a diameter of 10.6 meters. Inside, enormous concrete columns stand like towering giants. It's truly amazing!

把土地还给河流
Restoring Farmland to Rivers

离开日本的地下宫殿，水滴爸爸带领兄妹俩在时空中自由地穿梭。

After leaving the Underground Temple in Japan, Waterdrop Dad whisked his two kids away on another thrilling adventure, soaring through time and space.

原来，水滴家族穿越回 21 世纪了。在这个时期，各国尝试将“蓝绿空间”纳入防洪规划中。荷兰人也开始为河流创造更多的空间，恢复河流的蓄洪和生态功能。

The Waterdrop family has traveled back to the 21st century! During this time, many countries started thinking about how to include "blue and green spaces" in their flood control plans. In the Netherlands, they've been working on giving rivers more space to breathe, allowing them to store floodwater and support the local ecosystem.

荷兰城市奈梅亨的河道原状（河流处在高水位）
The river's original state in Nijmegen, Netherlands (at high water levels)

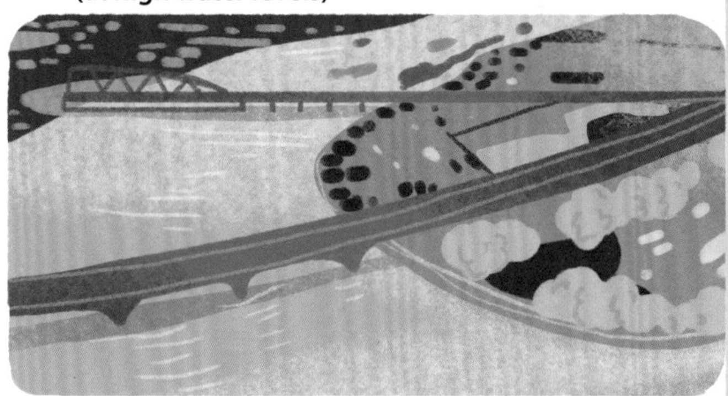

为河流创造更多空间
Giving the river more space

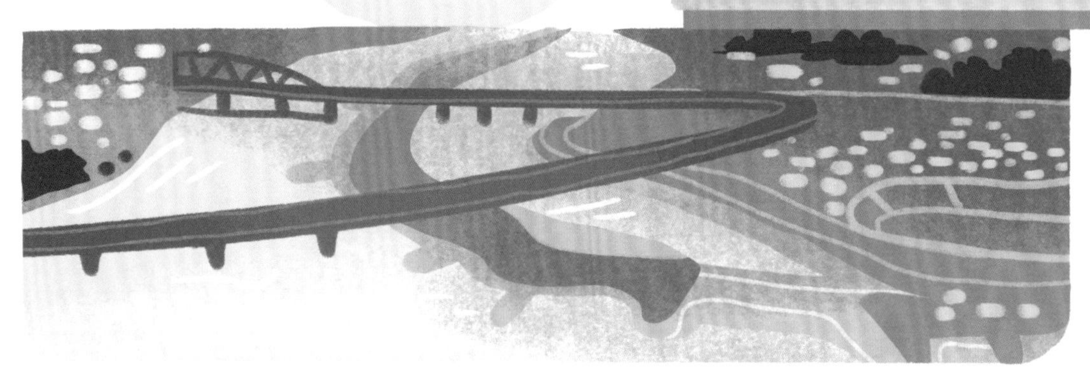

什么是“蓝绿空间”？
What is "blue and green space"?

“蓝绿空间”以水体、水道和植被为代表，指由河湖水系构成的蓝色空间和绿地系统构成的绿色空间。

"Blue and Green Space" refers to a natural environment consisting of water bodies, waterways, and vegetation, with the "blue" space made up of river and lake systems, and the "green" space formed by green areas such as parks and forests.

水滴家族话科普
The Waterdrop Family's Science Talk

水滴爸爸，人类治理洪水的观念是如何转变的呀？
Dad, how has the way people control floods changed?

问得好！随着全球气候的变化，极端天气事件频发。这让人类意识到，一味改造自然绝非长久之计。于是，“人类与洪水共存”成了21世纪的全新议题。
Great question! With climate change and more extreme weather happening around the world, people realized that trying to control nature all the time isn't the best approach. So now, in the 21st century, the idea of 'living with floods' has become a new way of thinking.

原来如此呀！
Ah, now I get it!

不怕淹的城市

Flood-Resilient Cities

一场强降雨来袭，欧洲多地遭遇水灾。还没来得及离开的水滴家族会遇到危险吗？

A heavy rainstorm hits, flooding many parts of Europe. Will the Waterdrop family, who haven't left yet, be in danger?

事实证明，有些城市并不怕淹。这是怎么做到的呢？

Surprisingly, some cities don't fear flooding. How do they manage it?

① 荷兰致力于打造“不怕淹的城市”，采用复合材料筑造坚固防潮的房屋。

In the Netherlands, cities are being designed to withstand floods. Builders use composite materials to create strong, moisture-proof homes.

坚固防潮的房屋
Strong, moisture-proof homes

② 在伦敦纽汉姆洪水高风险区，英国环保署建议：新建建筑物底层不得设计卧室，并需加装防洪屏障。

In London's Newham area, which is at high risk for flooding, the UK Environment Agency suggests that new buildings should not have bedrooms on the ground floor and should be equipped with extra flood barriers.

建筑底层不得设计卧室
No bedrooms on the ground floor

加装防洪屏障
Flood barriers added

③ 同时，伦敦还将周遭土地改造成湿地公园，有效规避暴雨和涨潮风险，“房”“洪”共融，为世界提供了“韧性城市”的先例。

Additionally, London has transformed surrounding areas into wetland parks. This helps prevent flooding from rainstorms and rising tides, setting an example for resilient cities.

改造而成的湿地公园
The transformed wetland park

水滴家族话科普
The Waterdrop Family's Science Talk

水滴爸爸，什么是韧性城市呀？
Dad, what is a resilient city?

问得好！韧性城市指的是在面临地震、洪水、火灾等突发事件时，能够快速响应，维持基本运转，并且可以在冲击结束后迅速恢复的城市。
Great question! A resilient city is one that can quickly respond to emergencies like earthquakes, floods, and fires. It can keep basic services running during the crisis and bounce back quickly after the disaster is over.

哇，真厉害！
Wow, that's impressive!

你知道吗？
Do You Know?

为了优化雨洪管理，美国提出了“低影响开发”的理念，中国则着力打造“海绵城市”，希望城市能够像海绵一样，在适应环境变化和应对自然灾害时具有良好的“韧性”。

To manage rain and floods better, the United States introduced the idea of "low-impact development", while China is working on creating "sponge cities". The idea is for cities to act like sponges, soaking up water and adapting to environmental changes, so they can better handle natural disasters.

海绵城市
Sponge city

洪水来临早知道

How Do We Predict Floods?

离开欧洲，时空飞船带着水滴家族又来到了中国。嘀嘀——突然，飞船响起了警报声！

Leaving Europe, the Time-Space Ship whisked the Waterdrop family off to China. Suddenly—beep beep—an alarm blared!

水滴爸爸带大家来到远程控制大厅，了解洪水预报预警系统的发展历程。

Waterdrop Dad led the family to the control room, where they learned about the history of flood forecasting and warning systems.

① 早在 1850 年，法国工程师尤金·贝尔格朗德创立了最早的洪水预报系统。

Back in 1850, French engineer Eugene Belgrand created the world's first flood forecasting system.

② 随着科技的进步，洪水预报的数据来源不再只有雨量计、水文站网，也需要雷达、遥感等预报技术。

As technology advanced, flood forecasting began using more than just rain gauges and hydrology networks. Today, it also relies on radar, remote sensing, and other high-tech forecasting tools.

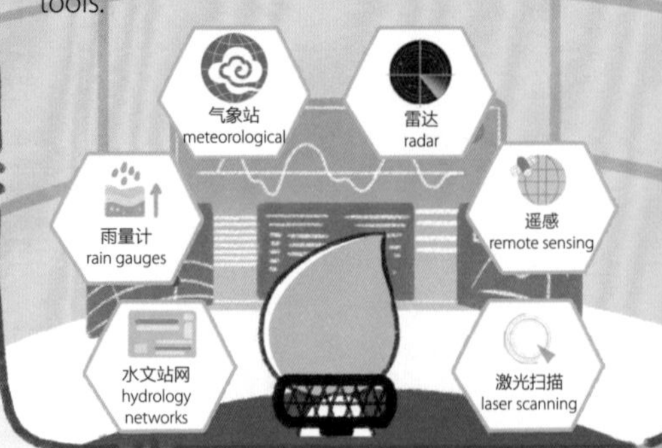

水滴家族话科普
The Waterdrop Family's Science Talk

水滴爸爸，人们常说的“四预建设”是什么呀？
Dad, what exactly are the “Four Pre-flood Efforts” ?

在水利管理中，预报、预警、预演、预案四个方面的能力建设，被统称为“四预建设”。
The “Four Pre-flood Efforts” are four important steps for flood management: forecasting, warning, practice drills, and planning. Together, they help us prepare for floods and reduce their damage.

我懂了。防洪减灾要做好，“四预建设”少不了！
Ah, I get it now! To protect against floods, the ‘Four Pre-flood Efforts’ are super important!

你听说过“洪水保险”吗？
Have you ever heard of "flood insurance"?

20 世纪 60—70 年代，美国出台了《全国洪水保险法》，对洪水灾害引起的经济损失给予经济赔偿。

In the 1960s and 1970s, the United States introduced *the National Flood Insurance Act*. This law helps people get financial support to recover from flood damage.

③ 终端预警也由口口相传、锣鼓、广播等形式，发展为平台、电话、短信、社交媒体甚至入户预警等。手机实时预警，将会成为未来的新趋势哦！

Flood warnings used to be spread by word of mouth, gongs, or radio. But now, we get warnings through platforms, phone calls, text messages, social media, and even alerts sent right to your home! Real-time mobile alerts are becoming the way of the future!

④ 如今，数字孪生技术赋能洪水预警预报。大数据及人工智能技术的加持，将进一步助力智慧水利建设！

Today, digital twin technology is enhancing flood forecasting and early warning systems. With the help of big data and artificial intelligence, it is set to play a key role in advancing smart water management!

洪灾之前做准备
Getting Ready Before a Flood

自从收到洪水预报，水滴妹妹就紧张得坐立不安，小水滴也有些手足无措。

Since they got the flood warning, Waterdrop Sister was fidgeting nervously, and Little Waterdrop didn't know what to do either.

水滴爸爸，我们该怎么办？
Dad, what should we do?

应对洪水有办法，做好准备不惊慌！
As long as we're prepared, there's no need to panic!

洪水来临前应该做哪些准备呢？水滴爸爸把经验传授给了兄妹俩。

So, what should we do to get ready? Waterdrop Dad began sharing his tips with the kids.

如何妥善安置贵重物品？
How to Keep Valuables Safe?

你可以将不便携带的贵重物品防水捆扎后埋入地下或放在高处，而现金和首饰等小件物品则可缝在衣物内随身携带。

You can protect things that are hard to carry by wrapping them in waterproof material and either burying them underground or putting them on a high shelf. For small things like jewelry and tickets, sew them into your clothes and keep them with you.

① 随时关注洪水预警信息，提前熟悉安全避险路线。

Stay updated on flood warnings and familiarize yourself with escape routes in advance.

② 备足食品、衣物、饮用水、生活用品和必要的医疗用品，并妥善安置贵重物品。

Prepare enough food, clothing, drinking water, daily necessities, and essential medical supplies. Store your valuables safely.

水滴家族话科普

The Waterdrop Family's Science Talk

水滴爸爸，除了准备应急物品，我们还能制作一些救生装置吗？

Dad, besides gathering emergency supplies, can we make any life-saving devices?

当然！我们可以搜集木盆、木桶、大件泡沫塑料等适合漂浮的材料，将它们捆扎在一起，加工成救生装置。

Of course! We can gather materials that float, like wooden basins, barrels, large foam pieces, and tie them together to create life-saving devices.

哇，这个办法真不错！

Wow, that sounds like a great idea!

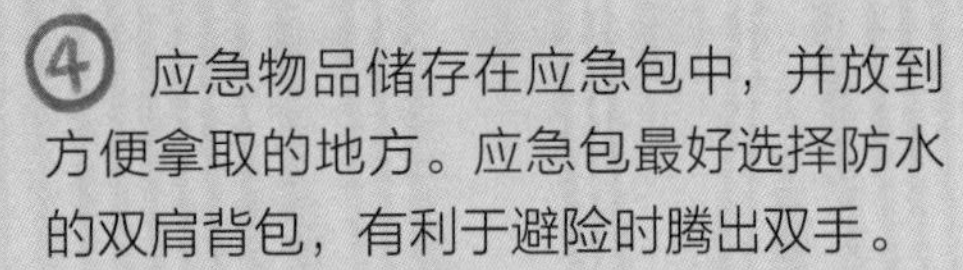

③ 准备可以发出求救信号的物品，比如手电、哨子、镜子、打火机、鲜艳的衣物等。被困时，吹哨、闪烁灯光比呼喊求救更有效。

Prepare items to signal for help, such as a flashlight, whistle, mirror, lighter, or brightly colored clothing. When trapped, using a whistle or flashing lights is more effective than shouting for help.

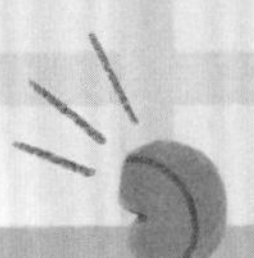

④ 应急物品储存在应急包中，并放到方便拿取的地方。应急包最好选择防水的双肩背包，有利于避险时腾出双手。

Store emergency supplies in an easy-to-reach emergency kit. A waterproof backpack is ideal as it keeps your hands free during evacuation.

洪水来时怎么办？

What to Do When a Flood Hits?

不久，汹涌的洪水如约而至，奔泻而来！这回，水滴家族会如何应对呢？

Before long, the roaring flood came rushing in, just as expected! How would the Waterdrop family handle it?

在水滴爸爸的指导下，小水滴和水滴妹妹学到了许多安全避险小常识。

With Dad's guidance, Little Waterdrop and Waterdrop Sister learned many useful tips on staying safe and finding shelter.

① 在时间充裕时，应按原定的避险路线，有组织地向山坡、高地转移。

If you have plenty of time, follow your planned escape route and move calmly to higher ground or a safe hillside.

② 在来不及转移时，可以爬到屋顶、大树、高墙上暂时避险，等待救援。

If there's no time to escape, find a safe spot like a roof, tree, or high wall to climb and wait for help.

你知道吗？
Do You Know?

注意！洪水发生时，不要在车里避雨；不要在地势低的地方逗留；发现高压线塔倾倒、电线低垂或断裂时，要远离避险。

Important! When a flood happens: never take shelter in your car; Don't stay in low-lying areas. If you see a power pole toppled, or wires sagging or broken, stay far away to stay safe.

③ 被洪水包围时，应利用身边的救生装置，如船只、木排、门板、木床等，做水上转移。

If the flood surrounds you, use anything around you that floats, like a boat, wooden raft, door, or even a wooden bed, to move safely across the water.

④ 有通信条件时，可向当地政府和防汛部门寻求救援；无通信条件时，可制造烟火、挥动鲜艳的衣物或集体同声呼救，向外界发出求助信号。

If you can communicate, call local authorities or the flood control team for help. If you can't, make signals with smoke, bright clothing, or shout together to ask for help.

漂浮建筑真奇妙

Amazing Floating Buildings

洪水终于退去，水滴家族重新返回时空飞船，愉快地聊起了天。

As the flood finally receded, the Waterdrop family returned to their Time-Space Ship, chatting happily as they went.

近年来，全球极端暴雨天气导致洪水频发，海平面上升。

Lately, we've seen more and more floods because of heavy rainfall, and sea levels are rising.

如果未来的地球变成了水的世界，该怎么办？

What if one day, the Earth becomes a water world?

问得好！我带你们去参观一种奇妙的建筑。

Great question! Let's go see something truly amazing.

水滴爸爸驾驶着飞船，带领兄妹俩参观漂浮建筑。

Waterdrop Dad steered the ship toward a special place – a floating building.

漂浮建筑在一定的条件下可以随水位涨落浮动，不易受洪水侵袭。这种巧妙的设计确实令人赞叹！

These incredible buildings can rise and fall with the water, so they're less likely to be damaged by floods. What a clever design!

在尼日利亚的贫民窟马可可，每年有长达四个月的雨季。原有的小学因洪水而不复存在。

In the slums of Makoko, Nigeria, the rainy season lasts for four months each year, causing devastating floods. The original primary school was swept away by the waters.

建筑师使用木材和竹子设计了一座水上学校。金字塔形结构使之稳定又便于漂浮，成为当地唯一一所学校。

To solve this, architects designed a floating school using wood and bamboo. Its pyramid-shaped structure is both stable and buoyant, making it the only school of its kind in the area.

韩国釜山计划建成全球第一座漂浮城市。

城市呈群岛状，居民将利用太阳能发电，食物和淡水自给自足。

Meanwhile, in Busan, South Korea, plans are underway to build the world's first floating city. This city will consist of a series of islands, where residents will harness solar power, grow their own food, and rely on their own fresh water supply.

洪水，让我们更好地共生共存吧！

Let's Live in Harmony with Floods!

暑假即将结束，水滴家族的亲子旅行也进入了尾声。

嗖——电光火石间，时空飞船带领大家回到了最初的起点。

As the summer vacation came to a close, so did the Waterdrop family's adventure.

Whoosh—in the blink of an eye, the Time-Space Ship zoomed them back to where it all began.

经过这次旅行，小水滴和水滴妹妹收获满满！

几千年来，人类与洪水的共存方式不断演进，从未知到可知，从除水害到兴水利，从驯服洪水到与河流和谐共生。

After their journey, Little Waterdrop and his sister had learned so much!

Over thousands of years, the way humans have lived with floods has changed. What was once a mystery is now understood. We've moved from simply protecting ourselves from floods to harnessing water for good, and from controlling floods to living peacefully alongside rivers.

水滴家族话科普

The Waterdrop Family's Science Talk

我发现，仅仅对抗洪水已不再是屡试不爽的方法！
I've realized that simply fighting floods is no longer a foolproof solution!

没错。时至今日，人类唯有秉持道法自然、天人合一的理念，才能在未来走上韧性防洪、可持续发展的道路。
Exactly. Today, humanity must embrace the idea of living in harmony with nature and following its natural laws in order to build resilient flood control systems and pursue sustainable development in the future.

看来，在不断变化的世界中，人类也要不断地努力创新才行！
It seems that in this ever-changing world, we must keep innovating to keep up!

人类与千年洪水

Humans and the Floods Across Millennia

（汉英对照）